Planet Earth

Written by Ian James
Illustrated by Andrew Farmer

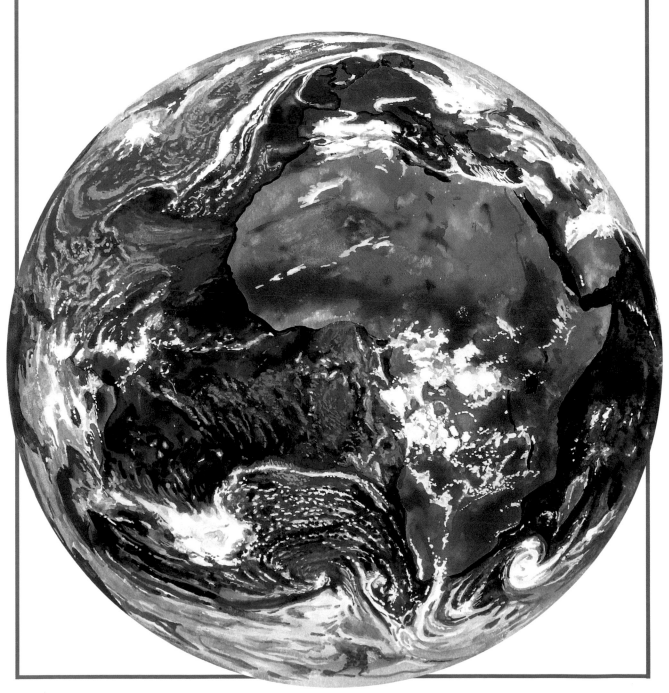

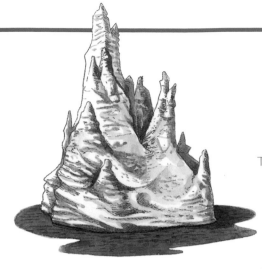

This is a Parragon Book
This edition published in 2001

Parragon
Queen Street House
4 Queen Street
Bath BA1 1HE, UK

ISBN 0 75254 823 9

Printed in Italy

Produced by Miles Kelly Publishing Ltd
Unit 11
Bardfield Centre
Great Bardfield
Essex
CM7 4SL

Contents

Archean 4,600–2,500 m.y.a.

Earth's formation

The geological time scale

The Earth was formed from a cloud of gas and dust around 4,600 million years ago.

What was the Earth like after it formed?

The Earth's surface was probably molten (hot and liquid) for many millions of years after its formation. The oldest known rocks are about 3,960 million years old.

When did living things first appear on Earth?

The oldest known fossils (of microscopic bacteria) are around 3,500 million years old. Primitive life forms may have first appeared on Earth about 3,850 million years ago.

Proterozoic 2,500–590 m.y.a.

The Archean and Proterozoic eons together occupied 87% of Earth history.

Why is the Cambrian period important?

During the Precambrian, most living creatures were soft-bodied and they left few fossils. During the Cambrian period, many creatures had hard parts, which were preserved as fossils in layers of rock.

What were the first animals with backbones?

Jawless fishes were the first animals with backbones. They appeared during the Ordovician period. Fishes with skeletons of cartilage, such as sharks, first appeared in the Devonian period.

When did plants start to grow on land?

The first land plants appeared in the Silurian period. Plants produced oxygen and provided food for the first land animals, amphibians. Amphibians first appeared in the Devonian period.

Why did the dinosaurs become extinct?

THE DINOSAURS FIRST APPEARED ON EARTH DURING THE TRIASSIC PERIOD. They became the dominant animals during the Jurassic period, but at the end of the Cretaceous period, 65 million years ago, they became extinct. Scientists still argue about why they disappeared. But many experts now believe that around 65 million years ago an enormous asteroid struck the Earth. The impact threw up a huge cloud of dust, which blocked out the sunlight for a long time. Land plants died and so the dinosaurs starved to death.

When did mammals first appear?

Mammals lived on Earth from at least the start of the Jurassic period. But they did not become common until after the extinction of the dinosaurs.

When did people first live on Earth?

Hominids (ape-like creatures that walked upright) first appeared on Earth more than four million years ago. But modern humans appeared only around 100,000 years ago.

The last 590 million years of Earth history are divided into eras and periods. 'M.y.a.' on the diagram means 'millions of years ago'.

How is Earth's history divided up?

SCIENTISTS DIVIDE THE LAST 590 MILLION YEARS OF EARTH'S HISTORY INTO THREE MAIN ERAS: the Paleozoic (a word meaning 'old life') era, the Mesozoic ('middle life') era, and the Cenozoic ('new life') era. The eras are then subdivided into periods – and some periods are further divided into epochs. The first period in the Paleozoic era is the Cambrian period. All of Earth's history before the Cambrian period is called the Precambrian. Scientists divide the Precambrian into two eons: the Archean and the Proterozoic. Scientists know little about life in the Precambrian, because fossils from that time are rare.

590 m.y.a.
505 m.y.a.
438 m.y.a.
408 m.y.a.
360 m.y.a.
286 m.y.a.
248 m.y.a.
213 m.y.a.
144 m.y.a.
65 m.y.a.
2 m.y.a.
Today

Precambrian
Cambrian period
Ordovician period
Silurian period
Devonian period
Carboniferous period
Permian period
Triassic period
Jurassic period
Cretaceous period
Tertiary period
Quaternary period

Paleozoic era
Mesozoic era
Cenozoic era

What are plates?

The Earth's hard outer layers are divided into large blocks called plates. Currents in the partly molten rocks inside the Earth slowly move the plates around.

How big are the plates?

The Earth's outer layers are split into seven large plates and about 20 small ones. The plates are about 70 to 100 km (43 to 62 miles) deep.

How the continents have drifted apart in the last 200 million years

Around 280 million years ago, the world's land areas moved together to form one supercontinent called Pangaea.

What happens when plates collide?

Along deep ocean trenches, one plate is pulled beneath another. There it is melted and destroyed. When continents collide, their edges are squeezed up into new mountain ranges.

How fast do plates move?

Plates move, on average, between 1 and 10 cm (0.4 and 4 in) a year. This may sound slow. But over millions of years, these small plate movements dramatically change the face of the Earth.

How do plates move apart?

PLATES CONSIST OF THE EARTH'S CRUST AND THE TOP PART OF THE MANTLE. The plates float on a partly molten layer within the mantle. Huge underwater mountain ranges, called ocean ridges, rise from the ocean bed. Along the middle of these ridges are valleys, where plates are being pulled apart by currents in the partly molten rocks below. As plates move apart, liquid rock, called magma, rises and plugs the gaps. When the magma hardens, it forms new crustal rock.

Who first suggested the idea of continental drift?

In the early 1800s, an American, F.B. Taylor, and a German, Alfred Wegener, both suggested the idea of continental drift. But scientists could not explain how the plates moved until the 1960s, following studies of the ocean floor.

Have fossils helped to prove continental drift?

Fossils of animals that could not possibly have swum across oceans have been found in different continents. This suggests that the continents were once all joined together and the animals could walk from one continent to another.

How plates change the face of the Earth

The ocean floor has huge ridges, where plates are moving apart. The gaps are filled with rising magma.

Plates consist of the Earth's crust and the rigid upper layer of the mantle.

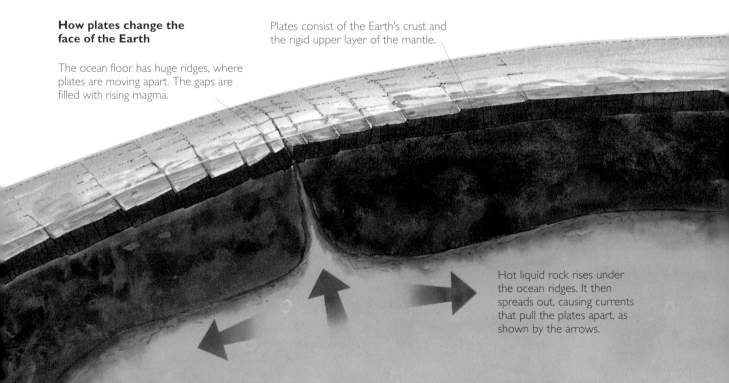

Hot liquid rock rises under the ocean ridges. It then spreads out, causing currents that pull the plates apart, as shown by the arrows.

How do volcanic islands form in the middle of oceans?

Volcanic islands form when magma rises from the mantle. Lava (the name for magma when it reaches the surface) piles up until it emerges above sea level.

Pangaea began to break apart around 180 million years ago.

Has the Earth always looked the same?

IF ALIENS HAD VISITED EARTH 200 MILLION YEARS AGO, THEY WOULD HAVE SEEN only one huge continent, called Pangaea, surrounded by one ocean. Around 180 million years ago, Pangaea began to break up. By 135 million years ago, a plate bearing South America was drifting away from Africa, creating the South Atlantic Ocean. By 100 million years ago, plates supporting India, Australia and Antarctica were also drifting away from Africa, and North America was moving away from Europe.

Can plates move sideways?

Plates not only move apart or push against each other, they can also move sideways along huge cracks in the ground called transform faults.

Around 65 million years ago, the Atlantic Ocean was opening up and India was moving towards Asia.

The map shows our world today, but plate movements are still changing the face of our planet.

Along the deep ocean trenches, ocean plates are pushed beneath other plates. Here, the plate supports a continent.

The rocks on the edge of the continents are folded by the pressure of plate movements.

Magma from the melted plate rises. Some emerges through volcanoes.

When plates move, the land is shaken by earthquakes.

The edge of the descending plate is melted, producing huge pockets of magma.

What earthquake in modern times caused the most damage?

In 1923, an earthquake struck Tokyo, capital of Japan. About 575,000 homes were destroyed in Tokyo and nearby Yokohama. About 142,800 people died.

Severe earthquakes occur around the edges of the plates that form the outer, rigid layers of the Earth.

Can scientists predict earthquakes?

In 1975, Chinese scientists correctly predicted an earthquake and saved the lives of many people. But scientists have not yet found any sure way of forecasting earthquakes.

Can animals sense when an earthquake is about to happen?

Scientists have noticed that animals often behave strangely before an earthquake. Horses rear up, dogs bark and snakes come out of their holes in the ground.

What is the San Andreas fault?

The San Andreas fault is a long transform fault in California. Movements along this plate edge have caused great earthquakes in San Francisco and Los Angeles.

Where are earthquakes likely to happen?

EARTHQUAKES CAN OCCUR ANYWHERE, WHENEVER ROCKS MOVE ALONG FAULTS (cracks) in the ground. But the most violent earthquakes occur most often around the edges of the plates that make up the Earth's hard outer layers. Plates do not move smoothly. For most of the time, their edges are jammed together. But gradually the currents under the plates build up increasing pressure. Finally the plates move in a sharp jerk. This sudden movement shakes all the rocks around it, setting off an earthquake.

Many earthquakes occur along transform faults

Severe earthquakes occur along ocean ridges and near ocean trenches. They also occur near another kind of plate edge called a transform fault.

Transform faults are long fractures in the Earth where plates move alongside each other.

What instruments record earthquakes?

Seismographs are sensitive instruments that record earthquakes. The shaking of the ground is recorded by a pen that marks the movements on a revolving drum.

Do earthquakes and volcanoes occur in the same places?

Most active volcanoes occur near the edges of plates that are moving apart and also where they are colliding. Earthquakes are common in these regions too.

How do earthquakes cause damage?

POWERFUL EARTHQUAKES SHAKE THE GROUND. THEY MAKE BUILDINGS SWAY and wobble until they collapse. The shaking sometimes breaks gas pipes or causes electrical short-circuits, starting fires. Earthquakes on high mountain slopes cause landslides that sometimes destroy towns in the valleys below. Earthquakes on the sea-bed trigger off waves called tsunamis. Tsunamis travel through the water at up to 800 kph (500 mph). As they approach land, the water piles up into waves many metres high. These waves cause great damage and loss of life.

Vibrations occur when the plates move, causing violent shaking of the ground. This destroys buildings and other structures, often with loss of life.

Pressure builds up until finally the rocks break and the plates move suddenly forwards.

Transform faults have ragged edges and, for most of the time, the plates are locked together.

When volcanoes erupt, they may hurl rocks and ash into the air, and lava may flow down slopes.

What makes volcanoes erupt?

VOLCANOES ERUPT WHEN HOT MOLTEN ROCK FROM DEEP DOWN IN EARTH'S mantle rises through the Earth's hard outer layers and reaches the surface. The molten rock is called magma, but when it reaches the surface, it is called lava. Most volcanoes occur near the edges of plates. Many rise along the ocean ridges where magma rushes up to fill the gaps formed as plates move apart. Other volcanoes get their magma from the plates that are melted as they are pulled beneath other plates.

Magma reaches the surface through vents, which are holes in the ground.

Lava may burst from a central vent or through side vents.

Lava flows burn everything in their paths.

Clouds of ash often block out the Sun. Ash falls on the land and it may bury towns. Mud flows occur when rain turns the ash into torrents of mud.

What are volcanoes made of?

Some volcanoes are cone-shaped and made of volcanic ash or cinders. Dome-shaped shield volcanoes are made of hardened lava. Intermediate volcanoes contain layers of both ash and lava.

Kinds of volcanoes
Shield volcano

Some volcanoes, shaped like upturned shields, are formed by 'quiet eruptions', in which long streams of very fluid lava are emitted.

What is a dormant volcano?

Some volcanoes erupt continuously for long periods. But other active volcanoes erupt only now and then. When they are not erupting, they are said to be dormant, or sleeping.

What is an extinct volcano?

Volcanoes that have not erupted in historic times are said to be extinct. This means that they are not expected to erupt ever again.

Do volcanoes do any good?

Volcanic eruptions cause tremendous damage, but soil formed from volcanic ash is fertile. Volcanic rocks are also used in building and chemical industries.

What are 'hot spots'?

Some volcanoes lie far from plate edges. They form over 'hot spots' – areas of great heat in the Earth's mantle. Hawaii in the Pacific Ocean is over a hot spot.

Explosive volcano

Explosive eruptions occur when the magma is thick and contains explosive gases. Explosive volcanoes are made of ash and cinders and are steep-sided.

Do all volcanoes erupt in the same way?

NO, THEY DON'T. VOLCANOES CAN EXPLODE UPWARDS OR SIDEWAYS, OR erupt 'quietly'. Trapped inside the magma in explosive volcanoes are lots of gases and water vapour. These gases shatter the magma and hurl columns of volcanic ash and fine volcanic dust into the air. Fragments of shattered magma are called pyroclasts. Sometimes, clouds of ash and hot gases are shot sideways out of volcanoes. They pour downhill at great speeds destroying everything in their paths. In 'quietly' erupting volcanoes the magma emerges on the surface as runny lava and flows downhill.

Intermediate volcano

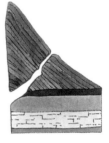

Intermediate volcanoes are cone-shaped. They are composed of alternating layers of ash and hardened lava.

What are hot springs and geysers?

Magma heats water in the ground. The hot water often bubbles up in hot springs. Sometimes, boiling water and steam are hurled into the air through geysers.

When lava and volcanic ash harden, they slowly break down to form soil.

What are the most valuable minerals?

Gemstones such as diamonds, rubies, sapphires and emeralds are valuable minerals. Gold and silver are also regarded as minerals, although they occur as native, or free, elements.

Ever since the Stone Age, people have used gemstones to make jewellery. Beautiful jewellery commands high prices.

Gemstones

Can minerals make you invisible?

People once believed in many superstitions about minerals. In the Middle Ages, people thought that you would become invisible if you wore an opal wrapped in a bay leaf.

What is the hardest mineral?

Diamond, a pure but rare form of carbon, is formed under great pressure deep inside the Earth. It is the hardest natural substance.

What are the most common rocks?

Sedimentary rocks cover 75% of the Earth's land surface. But igneous and metamorphic rocks make up 95% of the rocks in the top 16 km (10 miles) of the Earth's crust.

What are birthstones?

Birthstones are minerals that symbolize the month of a person's birth. For example, garnet is the birthstone for January, while ruby is the stone for people born in July.

What are the three main kinds of rock?

THERE ARE IGNEOUS, SEDIMENTARY AND METAMORPHIC ROCKS. IGNEOUS ROCKS are formed from cooled magma. Sometimes it cools on the surface to form such rocks as basalt. Other magma cools underground to create rocks called granites. Many sedimentary rocks are made from worn fragments of other rocks. For example, sandstone is formed from sand. Sand consists mainly of quartz, a mineral found in granite. And some limestones are made from the shells of sea creatures. Metamorphic rocks are rocks changed by heat and pressure. For example, great heat turns limestone into marble.

Fragments of sand, silt and mud are washed into lakes and seas. There they pile up in layers that harden into sedimentary rocks.

What are elements and minerals?

EARTH'S CRUST CONTAINS 92 ELEMENTS. THE TWO MOST COMMON ELEMENTS are oxygen and silicon. Also common are aluminium, iron, calcium, sodium, potassium and magnesium. These eight elements make up 98.59% of the weight of the Earth's crust. Some elements, such as gold, occur in a pure state. But most minerals are chemical combinations of elements. For example, minerals made of oxygen and silicon, often with small amounts of other elements, are called silicates. They include feldspar, quartz and mica – all found in granite.

Is coal a rock?

No. Rocks are inorganic (lifeless) substances. But coal, like oil and natural gas, was formed millions of years ago from the remains of once-living things. That is why coal, oil and gas are called fossil fuels.

Are some minerals more plentiful than others?

Many useful minerals are abundant. Other less common, but important minerals are in short supply and are therefore often recycled from scrap. Recycling saves energy, which has to be used to process metal ores.

What common rocks are used for buildings?

Two sedimentary rocks, limestone and sandstone, and the igneous rock granite are all good building stones. The metamorphic rock marble is often used to decorate buildings.

Earth movements and great heat turn igneous and sedimentary rocks into metamorphic rocks.

Igneous rocks are formed from magma, which may solidify beneath or on the Earth's surface. Surface rocks are constantly worn away by erosion.

How are fossils turned to stone?

When tree trunks or bones are buried, minerals deposited from water sometimes replace the original material. The wood or bone is then petrified, or turned to stone.

Fossils of ammonites are common in rocks of the Mesozoic era. Ammonites were molluscs, related to squid.

Ammonite fossil

What is carbonization?

Leaves usually rot quickly after plants die. But sometimes they float to the bottom of lakes and are buried under fine mud. Sediments above and below the leaf are gradually compressed and hardened into sedimentary rocks. Over time, bacteria gradually change the chemistry of the leaf until only the carbon it contains remains. The shape of the leaf is preserved in the rock as a thin carbon smear. This process is called carbonization.

Fossil teeth

What are fossils?

FOSSILS ARE THE IMPRESSIONS OF ANCIENT LIFE PRESERVED IN ROCKS. For example, when dead creatures are buried on the sea floor, the soft parts rot away, but the hard parts remain. Later, the mud and sand on the sea-bed harden into rock. Water seeping through the rock dissolves the hard parts, forming fossil moulds. Minerals fill the moulds to create casts, which preserve the shapes of the hard parts. Other fossils include outlines of leaves turned to stone, and insects in amber.

What are trace fossils?

Trace fossils give information about animals that lived in ancient times. Animal burrows are sometimes preserved, giving scientists clues about the creatures that made them. Other trace fossils include footprints preserved in hardened mud and quickly buried under more mud.

Fossil footprints

Fossil footprints are preserved in rocks. They are uncovered when overlying rocks are worn away.

What is amber?

Amber is a hard substance formed from the sticky resin of trees. Tiny animals were sometimes trapped in the resin. Their bodies were preserved when the resin hardened.

Fossil spider

Many tiny creatures have been fossilized in amber. They are unusual fossils because they consist of the actual bodies of ancient creatures.

Carbonized leaves

What was Piltdown man?
Some bones, thought to be fossils of an early human ancestor, were discovered at Piltdown, England, in 1913. But Piltdown man was a fake. The skull was human, but the jawbone came from an orang utan.

How do you date fossils?
Sometimes, dead creatures are found buried under volcanic ash. The ash sometimes contains radioactive substances that scientists can date. Hence, they can work out the time when the animals lived.

Trilobite fossils

What can scientists learn from fossils?

FROM THE STUDY OF FOSSILS – KNOWN AS PALAEONTOLOGY – SCIENTISTS CAN learn about how living things evolved on Earth. Fossils can also help palaeontologists to date rocks. This is because some species lived for only a short period on Earth. So, if the fossils of these creatures are found in rocks in different places, the rocks must have been formed at the same time. Such fossils are called index fossils. Important index fossils include species of trilobites, graptolites, brachiopods, crinoids, ammonites and belemnites.

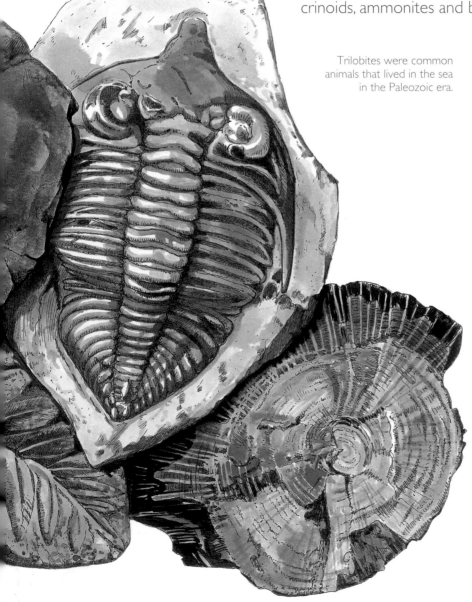

Trilobites were common animals that lived in the sea in the Paleozoic era.

Has flesh ever been preserved as a fossil?
In Siberia, woolly mammoths, which lived more than 40,000 years ago, sank in swampy ground. When the soil froze, their complete bodies were preserved in the icy subsoil.

What is eohippus?
Eohippus is the name of the dog-sized ancestor of the horse, which lived around 55 million years ago. Fossil studies of eohippus and its successors have shown how the modern horse evolved.

Petrified logs

Petrified logs were formed when water replaced the molecules in buried logs with minerals. Slowly, stone replicas of the logs were produced.

Limestone caves

Limestone caves are worn out by chemical weathering. They often contain stalactites and stalagmites.

What are stalagmites?

Stalagmites are the opposite of stalactites. They are columns of calcium carbonate deposited by dripping water. But stalagmites grow upwards from the floors of caves.

Does water react chemically with other rocks?

Water dissolves rock salt. It also reacts with some types of the hard rock granite, turning minerals in the rock into a clay called kaolin.

What are stalactites?

Water containing a lot of calcium carbonate drips down from the ceilings of limestone caves. The water gradually deposits calcium carbonate to form hanging, icicle-like structures called stalactites.

What is ground water?

Ground water is water that seeps slowly through rocks, such as sandstones and limestones. The top level of the water in the rocks is called the water table. Wells are dug down to the water table.

How quickly is the land worn away?

Scientists have worked out that an average of 3.5 cm (1.4 in) is worn away from land areas every 1,000 years. This sounds slow but, over millions of years, mountains are worn down to make plains.

What are pot-holers?

Pot-holes, or swallow holes, are holes in the ground where people called pot-holers can climb down to explore limestone caves. Pot-holers may face danger when sudden rains raise the water level in caves.

What are springs?

Springs occur when ground water flows on to the surface. Springs are the sources of many rivers. Hot springs occur in volcanic areas, where the ground water is heated by magma.

How does weathering help to shape the land?

WEATHERING IS THE BREAKDOWN AND DECAY OF ROCKS ON THE EARTH'S SURFACE. The wearing away of the rock limestone is an example of chemical weathering. Limestone consists mostly of the chemical calcium carbonate. This chemical reacts with rainwater containing carbon dioxide, which it has dissolved from the air. The rainwater is a weak acid that slowly dissolves the limestone. The rainwater opens up cracks in the surface, wearing out holes that eventually lead down to a maze of huge caves linked by tunnels.

Weathering of the land

Weathering is rapid on sloping land, where worn rocks tumble downhill.

Ground water flows out of limestone caves to form the source of a river.

Can the Sun's heat cause mechanical weathering?

In hot, dry regions, rocks are heated by the Sun, but they cool at night. These changes crack rock surfaces, which peel away like the layers of an onion.

Can plants change the land?

Plant roots can break up the rock. When the seed of a tree falls into a crack in a rock, it develops roots that push downwards. As the roots grow, they push against the sides of the crack until the rock splits apart.

What is biological weathering?

Biological weathering includes the splitting apart of rocks by tree roots, the breaking up of rocks by burrowing animals, and the work of bacteria, which also helps to weather rocks.

How does the action of frost break up rocks?

AT NIGHT IN THE MOUNTAINS, PEOPLE MAY HEAR SOUNDS LIKE GUNSHOTS. These sounds are made by rocks being split apart by frost action. Frost action, an example of mechanical weathering, occurs when water in cracks in rocks freezes and turns into ice. Ice takes up nearly one-tenth as much space again as water, and so it exerts pressure, widening the cracks until they split apart. On steep slopes, shattered rocks tumble downhill and pile up in heaps, called scree or talus.

Frost action affects high mountain slopes, where water freezes at night.

Surface water flows into layers of limestone and hollows out caves.

Worn rocks pile up in heaps called scree or talus.

The roots of trees can split rock apart.

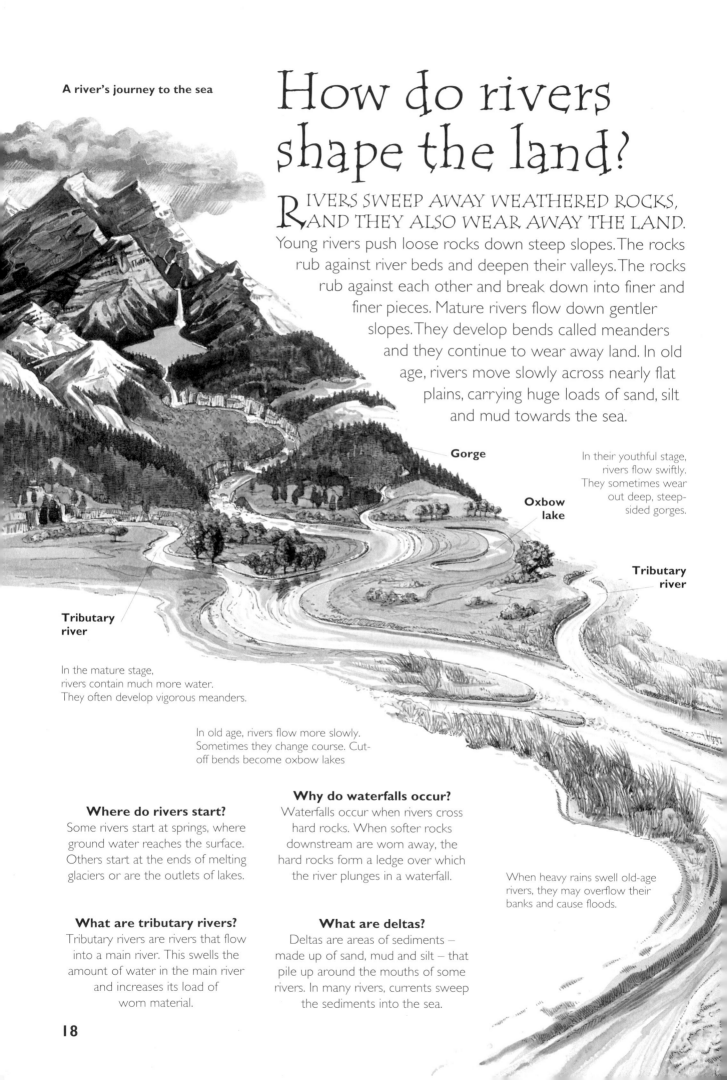

How do rivers shape the land?

RIVERS SWEEP AWAY WEATHERED ROCKS, AND THEY ALSO WEAR AWAY THE LAND. Young rivers push loose rocks down steep slopes. The rocks rub against river beds and deepen their valleys. The rocks rub against each other and break down into finer and finer pieces. Mature rivers flow down gentler slopes. They develop bends called meanders and they continue to wear away land. In old age, rivers move slowly across nearly flat plains, carrying huge loads of sand, silt and mud towards the sea.

Gorge

Oxbow lake

In their youthful stage, rivers flow swiftly. They sometimes wear out deep, steep-sided gorges.

Tributary river

Tributary river

In the mature stage, rivers contain much more water. They often develop vigorous meanders.

In old age, rivers flow more slowly. Sometimes they change course. Cut-off bends become oxbow lakes

Where do rivers start?
Some rivers start at springs, where ground water reaches the surface. Others start at the ends of melting glaciers or are the outlets of lakes.

Why do waterfalls occur?
Waterfalls occur when rivers cross hard rocks. When softer rocks downstream are worn away, the hard rocks form a ledge over which the river plunges in a waterfall.

What are tributary rivers?
Tributary rivers are rivers that flow into a main river. This swells the amount of water in the main river and increases its load of worn material.

What are deltas?
Deltas are areas of sediments – made up of sand, mud and silt – that pile up around the mouths of some rivers. In many rivers, currents sweep the sediments into the sea.

When heavy rains swell old-age rivers, they may overflow their banks and cause floods.

What are spits?

Waves and currents transport sediments along coasts. In places where the coasts change direction, the worn sand and pebbles pile up in narrow ridges called spits.

How can people slow down wave erosion?

Along the beaches at many coastal resorts, walls are built at right angles to the shore. These walls, called groynes, slow down the movement of sand on the beaches by waves and sea currents.

What is a baymouth bar?

Some spits join one headland to another. They are called baymouth bars, because they cut off bays from the sea, turning them into lagoons.

Does the sea wear away the land?

Waves wear away soft rocks to form bays, while harder rocks on either side form headlands. Parts of the coast of Northeast England have been worn back by up to 5 km (3 miles) since the days when the Romans ruled the area.

Caves, arches and stacks

Waves hollow out caves in rocky headlands. Blow-holes form above the caves.

When two caves in a headland meet, a natural arch occurs.

When a natural arch collapses, the tip of the headland becomes an isolated stack.

Can sea waves shape coasts?

LARGE STORM WAVES BATTER THE SHORE. THE WAVES PICK UP SAND AND PEBBLES, hurling them at cliffs. This hollows out the bottom layers of the cliff until the top collapses and the cliff retreats. Waves hollow out bays in soft rocks, leaving hard rocks jutting into the sea as headlands. Waves then attack the headlands from both sides, wearing out caves. When two caves meet, a natural arch is formed. When the arch collapses, all that remains is an isolated rock, called a stack.

Mud carried by a river is often dumped near the river's mouth to form large mud flats. At high tide the sea often covers these areas.

What are fiords?

Fiords are deep, steep-sided valleys that wind inland along coasts. They were once river valleys that were deepened by glaciers during the last ice age.

What are erratics?

Erratics are boulders made of a rock that is different from the rocks on which they rest. They were transported to their present positions by moving ice.

How much of the world is covered by ice?

Ice covers about 11% of the world's land area. But during the last ice age, it spread over much of northern North America and Europe. The same ice sheet reached what is now New York City in America, and covered London in England.

How does ice shape the land?

IN COLD MOUNTAIN AREAS, SNOW PILES UP IN HOLLOWS. GRADUALLY, THE SNOW becomes compacted into ice. Eventually, the ice spills out of the hollows and starts to move downhill to form a glacier. Glaciers are like conveyor belts. On the tops of glaciers are rocks shattered by frost action that have tumbled downhill. Other rocks are frozen into the sides and bottoms of glaciers. They give glaciers the power to wear away rocks and deepen the valleys through which they flow. Ice-worn valleys are U-shaped, with steep sides and flat bottoms. This distinguishes them from V-shaped river valleys.

A valley glacier

The ice spills downhill to form rivers of ice called glaciers. The glaciers carry much worn rock, called moraine.

At the end of the glacier, the ice melts, creating streams that sweep away the glacier's rocky load.

What are the world's largest bodies of ice?

The largest bodies of ice are the ice sheets of Antarctica and Greenland. Smaller ice caps occur in the Arctic, while mountain glaciers are found around the world.

The world map during the Ice Age

Ice covered much of northern North America, Europe and Asia during the Ice Age.

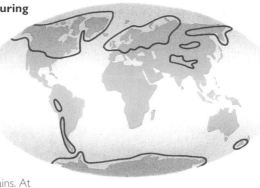

Snow falls on mountains. At the higher levels, the snow piles up year by year.

Snow in mountain basins, called cirques, becomes compressed into glacier ice.

What is an ice age?

During ice ages, average temperatures fall and ice sheets spread over large areas that were once ice-free. Several ice ages have occurred in Earth's history.

When did the last ice age take place?

The last ice age began about two million years ago and ended 10,000 years ago. The ice age included warm periods and long periods of bitter cold.

How can we tell that an area was once covered by ice?

Certain features in the landscape were made by ice during the ice ages. Mountain areas contain deep, steep-sided valleys that were worn out by glaciers. Armchair-shaped basins where glacier ice once formed are called cirques. Knife-edged ridges between cirques are called arêtes, while peaks called horns were carved when three or more cirques formed back-to-back. Boulders and other material carried by ice is called moraine. Moraine ridges show that ice sheets once reached that area.

How does wind-blown sand shape scenery?

IN DESERTS, WIND-BLOWN SAND IS IMPORTANT IN SHAPING THE SCENERY. Winds lift grains of sand, which are then blown and bounced forward. Sand grains are heavy and seldom rise over 2 metres (6 ft) above ground level. But, at low levels, wind-blown sand acts like the sand-blasters used to clean dirty city buildings. It also polishes rocks, hollows out caves in cliffs and undercuts boulders. Boulders whose bases have been worn by wind-blown sand are top-heavy and mushroom-shaped, perched on a narrow stem.

Desert scenery

What are wadis?
Wadis are dry waterways in deserts. Travellers sometimes shelter in them at night. But a freak storm can soon fill them with water and people sleeping in the wadis are in danger of being drowned.

Oases are places in deserts where water comes to the surface or where people can obtain water from wells.

Large areas of desert are covered with gravel and pebbles. These areas are called reg.

Can water change desert scenery?
Thousands of years ago, many deserts were rainy areas and many land features were shaped by rivers. Flash floods sometimes occur in deserts. They sweep away much worn material.

What are dust storms?
Desert winds sweep fine dust high into the air during choking dust storms. Wind from the Sahara in North Africa is often blown over southern Europe, carrying the pinkish dust with it.

What are the main types of desert scenery?
Arabic words are used for desert scenery. Erg is the name for sandy desert, reg is land covered with gravel and pebbles, and hammada is the word for areas of bare rock.

Why do people in deserts wear heavy clothes?
Deserts are often cold at night and heavy clothes keep people warm. Long cloaks and headdresses also help to keep out stinging wind-blown sand and dust and prevent sunburn.

Wind-blown sand is responsible for carving top-heavy mushroom rocks that stand on thin stems.

Mushroom rocks

Wind-blown sand erodes the bottoms of rocks, wearing them to a narrow stem.

Barchans

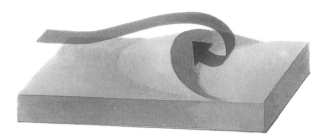

Crescent-shaped dunes form in sandy deserts where wind directions are constant.

How are sand dunes formed?

THE WIND BLOWING ACROSS DESERT SANDS PILES THE SAND UP IN HILLS CALLED DUNES. Where the wind directions keep changing, the dunes have no particular shape. But when they blow mainly from one direction, crescent-shaped dunes called barchans often form. Barchans may occur singly or in clusters. When winds drive sand dunes forward they form seif dunes, named after an Arabic word meaning sword. Sometimes, advancing dunes bury farmland. To stop their advance, people plant trees and grasses to anchor the sand.

What are oases?
Oases are places in deserts that have water supplies. Some oases have wells tapping ground water. Sometimes, the water bubbles up to the surface in a spring.

What is desertification?
Human misuse of the land near deserts, caused by cutting down trees and overgrazing grasslands, may turn fertile land into desert. This is called desertification. Natural climate changes may also create deserts. This happened in the Sahara around 7,000 years ago.

About one-fifth of the world's deserts are covered in sand (erg). There are also large areas of bare rock (hammada).

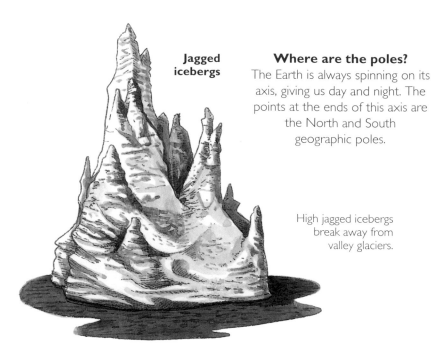

Jagged icebergs

Where are the poles?
The Earth is always spinning on its axis, giving us day and night. The points at the ends of this axis are the North and South geographic poles.

High jagged icebergs break away from valley glaciers.

Why are icebergs dangerous to shipping?
Icebergs are huge chunks of ice that break off from glaciers. They float in the sea with nine-tenths of their bulk submerged, which makes them extremely dangerous to shipping. Icebergs from Greenland have sunk ships off the coasts of North America.

What are magnetic poles?
The Earth is like a giant magnet, with two magnetic poles. They lie near the geographic poles, though their positions change from time to time.

Icebergs in the oceans

The bulk of the ice in icebergs is hidden beneath the waves.

What it is like around the North Pole?

IT IS BITTERLY COLD. THE NORTH POLE LIES IN THE MIDDLE OF THE ARCTIC OCEAN, which is surrounded by northern North America, Asia and Europe. Sea ice covers much of the ocean for most of the year. In spring, the sea ice is around 3 metres (9 ft) thick in mid-ocean and explorers can walk across it. The Arctic Ocean contains several islands, including Greenland, the world's largest. A huge ice sheet, the world's second largest, covers more than four-fifths of Greenland.

Is the ice around the poles melting?
In parts of Antarctica, the ice shelves began to melt in the 1990s. Some people think this shows that the world is getting warmer because of pollution.

Icebergs contain worn rocks that have been eroded from the land.

Icebergs melt as they float away from polar regions and the climate becomes warmer.

What animals live in polar regions?
Penguins are the best known animals of Antarctica. Polar bears, caribou, musk oxen and reindeer are large animals that live in the Arctic region.

As icebergs melt, the rocks in the ice sink down and settle on the ocean bed.

What are ice shelves?

Ice shelves are large blocks of ice joined to Antarctica's ice sheet but which jut out over the sea. When chunks break away, they form flat, table-topped icebergs. Some of them are huge. One covered an area about the size of Belgium.

Flat-topped icebergs form off the coast of Antarctica.

How thick is the ice in Antarctica?

THE SOUTH POLE LIES IN THE COLD AND WINDY CONTINENT OF ANTARCTICA, which is larger than either Europe or Australia. Ice and snow cover 98% of Antarctica, although some coastal areas and high peaks are ice-free. The Antarctic ice sheet is the world's largest, and contains about seven-tenths of the world's fresh water. In places, the ice is up to 4.8 km (3 miles) thick. The world's record lowest temperature, −89.2°C (−128.6°F), was recorded at the Vostok research station in 1983.

What is the Great Barrier Reef?

The Great Barrier Reef lies off the northeast coast of Australia. It is the world's longest group of coral reefs and islands. It is about 2,000 km (1,242 miles) long.

Where is the 'Great Pebble'?

The word Uluru is an Australian Aboriginal word meaning 'great pebble'. Also called Ayers Rock, it is the world's biggest monolith (single rock).

Grand Canyon

The Grand Canyon is regarded as one of the seven natural wonders of the world.

**Uluru
(Ayers Rock)**

Uluru, or Ayers Rock, is a tourist attraction in central Australia.

Where can you find 'smoke that thunders'?

The local name of the beautiful Victoria Falls on the River Zambezi between Zambia and Zimbabwe is Mosi oa Tunya. This name means 'smoke that thunders'.

Victoria Falls

The Victoria Falls in Africa was named after Queen Victoria by the missionary David Livingstone.

What are natural wonders?

THE ANCIENT GREEKS AND ROMANS MADE LISTS OF THE SEVEN WONDERS of the World. They were all made by people, but the Earth also has many natural wonders, created by the forces that continuously shape our planet. Most lists of natural wonders include the Grand Canyon in the southwestern United States. It is the world's largest canyon and the most awe-inspiring. The canyon is 446 km (277 miles) long and about 1.6 km (1 mile) deep. It was worn out by the Colorado River over the last six million years.

Where is the Matterhorn?

The Matterhorn is a magnificent mountain on Switzerland's border with Italy. It was created by glaciers wearing away the mountain from opposite sides.

What mountains look like dragon's teeth?

Steep-sided hills made of limestone, found around the town of Guilin in southeastern China, resemble rows of giant dragon's teeth. Rainwater wore out these strange-looking hills.

What is the world's mightiest river?

The Amazon River in South America contains far more water than any other river. Its river basin (the region it drains) is also the world's largest.

What is Meteor Crater?

Meteor Crater in Arizona, America, is a circular depression. It resembles a volcanic crater, but was formed about 50,000 years ago when a meteor hit our planet. The crater is 1,275 metres (4,180 ft) wide and 175 metres (570 ft) deep.

Where is the world's highest mountain range?

The Himalayan range in Asia contains 96 of the world's 109 peaks that are more than 7,315 metres (24,000 ft) above sea level. One of these peaks is Mount Everest, the world's highest mountain.

Where do the world's natural wonders occur?

THE WORLD'S NATURAL WONDERS CAN BE FOUND IN EVERY CONTINENT AND some, such as Australia's Great Barrier Reef, occur in the oceans. Many people now work to protect natural wonders so that they can be enjoyed by people in the future. One important step in protecting them was made in 1872, when the world's first national park was founded at Yellowstone, site of the famous geyser called Old Faithful, in the northwestern United States. Since then, national parks have been founded around the world.

What Japanese wonder attracts pilgrims?

Mount Fuji in Japan is a beautiful volcanic cone. Many people regard it as a sacred mountain – a dwelling place for the gods – and they make long pilgrimages to the top.

Limestone peaks

Near Guilin, China, are strange limestone hills that have become a major tourist attraction.

How can people protect natural wonders?

People enjoy visiting natural wonders, but they can damage the land, cause traffic pollution and spread litter. Many natural wonders are now protected by governments in areas called national parks.

Matterhorn

The Matterhorn reaches a height of 4,478 metres (14,692 ft) above sea level.

The Amazon

The Amazon drains a huge region that contains the world's largest rainforest.

What is air pollution?

Air pollution occurs when gases such as carbon dioxide are emitted into the air by factories, homes and offices. Vehicles also cause air pollution, which produces city smogs, acid rain and global warming.

What is coastal pollution?

Coral reefs and mangrove swamps are breeding places for many fishes. The destruction or pollution of these areas is threatening the numbers of fishes in the oceans.

Can deserts be farmed?

In Israel and other countries, barren deserts have been turned into farmland by irrigation. The land is watered from wells that tap ground water, or the water is piped from far-away areas.

Will global warming affect any island nations?

Coral islands are low-lying. If global warming melts the world's ice, then sea levels will rise. Such countries as the Maldives and Kiribati will vanish under the waves.

Can the pollution of rivers harm people?

When factories pump poisonous wastes into rivers, creatures living near the rivers' mouths, such as shellfish, absorb poison into their bodies. When people eat such creatures, they, too, are poisoned.

What is soil erosion?

Natural erosion, caused by running water, winds and other forces, is a slow process. Soil erosion occurs when people cut down trees and farm the land. Soil erosion on land made bare by people is a much faster process than natural erosion.

What happens when people exploit the Earth?

What is happening to the world's rainforests?

The rainforests in the tropics are being destroyed. These forests contain more than half of the world's living species. Many of them are now threatened with extinction. Huge forest fires in 1997 and 1998 destroyed large areas of rainforest.

Rainforest destruction

IN MANY AREAS, PEOPLE ARE CHANGING THE EARTH AND CAUSING GREAT HARM through pollution. They are cutting down forests to produce farmland. But in some places, rain and winds wear away newly exposed soil, causing soil erosion and making the land infertile. Factories, vehicles and homes burn fuels that release gases into the air. These so-called greenhouse gases trap heat. They cause global warming and change world climates. Other kinds of pollution include the poisoning of rivers and seas by factory and household wastes.

Rainforests are cut down by loggers who want to sell valuable hardwoods.

Land stripped bare of trees is exposed to the wind and the rain, which cause serious soil erosion.

How have people turned sea into land?

IN CROWDED COUNTRIES, PEOPLE SOMETIMES TURN USELESS COASTAL LAND INTO FERTILE FARMLAND. The Netherlands is a flat country and about two-fifths of it is below sea level at high tide. The Dutch have created new land by building dykes (sea walls) around areas once under the sea, called polders. Rainwater washes out the salt from the soil and the polder land finally becomes fertile. Global warming could affect the Netherlands. Increasingly stormy weather and rises in the sea level could cause massive flooding.

What is the biggest continent?

ASIA COVERS AN AREA OF 44,009,000 SQ KM (16,992,000 SQ MILES).
The other continents, in order of size, are Africa (30,246,000 sq km/11,678,000 sq miles), North America (24,219,000 sq km/9,351,000 sq miles), South America (17,832,000 sq km/6,885,000 sq miles), Antarctica (14,000,000 sq km/5,400,400 sq miles), Europe (10,443,000 sq km/4,032,000 sq miles), and Australia (7,713,000 sq km/2,978,000 sq miles).

What is the world's largest island?
Greenland covers about 2,175,000 sq km (840,000 sq miles). Geographers regard Australia as a continent and not as an island.

How much of the world is covered by land?
Land covers about 148,460,000 sq km (57,300,000 sq miles), or 29 % of the world's surface. Water covers the remaining 71 %.

What is the world's largest bay?
Hudson Bay in Canada covers an area of about 1,233,000 sq km (476,000 sq miles). It is linked to the North Atlantic Ocean by the Hudson Strait.

What is the world's largest high plateau?
The wind-swept Tibetan Plateau in China covers about 1,850,000 sq km (715,000 sq miles).

Where is the world's lowest point on land?
The shoreline of the Dead Sea, between Israel and Jordan, is 400 metres (1,312 ft) below the sea level of the nearby Mediterranean Sea.

What is the world's highest peak?
Mount Everest on Nepal's border with China reaches 8,848 metres (29,029 ft) above sea level. Measured from its base on the sea floor, Mauna Kea, Hawaii, is 10,203 metres (33,474 ft) high. But only 4,205 metres (13,796 ft) appear above sea level.

Earth seen from space

Viewed from space, we can see how small Planet Earth is in the setting of the Universe. Pictures like this have made people aware of the necessity of protecting our planet home.

Which is the deepest lake?
Lake Baikal, in Siberia, eastern Russia, is the world's deepest lake. The deepest spot measured so far is 1,637 metres (5,371 ft).

What is the largest inland body of water, or lake?
The salty Caspian Sea, which lies partly in Europe and partly in Asia, has an area of about 371,380 sq km (143,390 sq miles). The largest freshwater lake is Lake Superior, one of the Great Lakes of North America. Lake Superior has an area of 82,350 sq km (31,796 sq miles).

Is there a lake under Antarctica?
Scientists have found a lake, about the size of Lake Ontario in North America, hidden under Antarctica. It may contain creatures that lived on Earth millions of years ago.

What is the world's largest river basin?
The Amazon river basin in South America covers about 7,045,000 sq km (2,720,000 sq miles). The Madeira River, which flows into the Amazon, is the world's longest tributary, at 3,380 km (2,100 miles).

What is the world's longest river?
The Nile in north-eastern Africa is 6,617 km (4,112 miles) long. The second longest river, the Amazon in South America, discharges 60 times more water than the Nile.

What is the deepest cave?
The Réseau Jean Bernard in France is the deepest cave system. It reaches a depth of 1,602 metres (5,256 ft).

What is the world's largest desert?
The Sahara in North Africa covers an area of about 9,269,000 sq km (3,579,000 sq miles). This is nearly as big as the United States.

Where do most people live?

THE CONTINENT WITH THE LARGEST POPULATION IS ASIA, WHICH HAS more than 3,000 million people. Europe ranks second in world population, followed by Africa, North America, South America and Australia. The continent of Antarctica has no permanent population at all.

The patterns of cloud show how weather changes. But pollution is changing the climate and this may have serious effects on all living things on Planet Earth.

Index